A FRE

Attract Money Forever will deepen your understanding of metaphysics and mind-power principles as they relate to attracting money, manifesting abundance, and governing material reality. You'll learn how to use time-tested, time-honored, practical, and spiritual techniques to be more prosperous and improve your life in astounding and meaningful ways. Visit jamesgoijr.com/subscriber-page.html for your free download copy of this amazing book and to receive James's free monthly *Mind Power & Money Ezine*.

ADVANCED MANIFESTING MADE EASY

Books by James Goi Jr.

How to Attract Money Using Mind Power

Attract Money Forever

Ten Metaphysical Secrets of Manifesting Money

Also

Advanced Manifesting Made Easy

Aware Power Functioning

The God Function

And

My Song Lyrics (multiple volumes)

JGJ Thoughts (multiple volumes)

Note

James continues to write new books. To see a complete list, visit James on Amazon at Amazon.com/author/JamesGoiJr

ADVANCED MANIFESTING MADE EASY

— JGJ —

A METAPHYSICAL AND SPIRITUAL PATH TO ABUNDANCE, PROSPERITY, SUCCESS, ACHIEVEMENT, WEALTH, AND WISDOM

JAMES GOI JR.

JGJ
JAMES GOI JR.
LA MESA, CALIFORNIA

Copyright © 2017 by James Goi Jr.

All rights reserved. Brief passages of this book may be used in reviews but except as allowed by United States and international copyright law, no part of this book may be reproduced or transmitted in any form or by any means, electronic, mechanical, magnetic, photographic including photocopying, recording, or by any information storage and retrieval system without prior written permission of the publisher. No patent liability is assumed with respect to the use of the information contained herein, and the publisher and author assume no responsibility for errors, inaccuracies, omissions, or inconsistencies.

The information contained in this book is intended to be educational and not prescriptive, and is not meant to replace consultation with professionals in any field. Always seek professional help and guidance when needed.

ISBN:
978-1-68347-004-5 (Trade Paperback)
978-1-68347-005-2 (.mobi)

Published by:
James Goi Jr.
P.O. Box 563
La Mesa, CA, 91944
www.jamesgoijr.com

TRADEMARK NOTICE: The Attract Money Guru™ and Books to Awaken, Uplift, and Empower™ are trademarks of James Goi Jr.

CONTENTS

Preface

1. To Manifest and Create 1
2. Brushstrokes 5
3. No Brushstrokes? 9
4. The Foundation 13
5. It Already Is 17
6. Oneness 21
7. Scratching an Itch 25
8. Observation Equals Influence 29
9. A Great Responsibility 33
10. Point of View 37
11. Choose 41
12. In This Moment 45
13. The Reflection 49
14. How You See It 53

Afterword 57

About the Author 61

Special Acknowledgement 65

Further Reading 69

PREFACE

This is one of three books I wrote in 2010-2011 that sprang forth spontaneously when I sat down to write something much shorter and pretty much unrelated. After two days, and just two writing sessions, I had this book. (The other two books are *Aware Power Functioning* and *The God Function*.)

I had the distinct feeling while writing that the book had already been written somewhere else and that I was just putting it down on paper to bring it into the here and now. Other writers have talked of similar types of processes and experiences.

The book was not planned out or outlined, as is more or less standard practice in the writing of non-fiction books such as this, but instead was just written from one end to the other. In preparation for publishing this book, I separated it into titled chapters, did some editing and cleaning up, and the result is the book you now hold in your hands.

Even though this book "sprang forth spontaneously," there was something very different about the writing of it when compared with the experiences of writing others I've written in a similar way, and that includes *Aware Power Functioning* and *The God Function*.

With *this* book, right from the start, I grabbed my copy of the *Reader's Digest Oxford Complete Wordfinder* (dictionary and thesaurus), and I looked up the word *manifest*. I knew the meaning of the word as I was going to use it but felt compelled to look it up.

Then things started to get really interesting. Next, I found myself looking up the word *demonstrate*, a word I also used in the writing after finding it listed as a synonym for *manifest*. I knew the meaning of that word, too, but still just had to look it up.

Well, this process took hold of me, and I found myself looking up word after word after word. And the amazing thing to me was that this process of looking up the words was actually *guiding* my writing. It was strange and, really, otherworldly.

I don't know what else to say about all of this looking up of words other than to say that without having looked up a multitude of words and letting myself be guided by their meanings, their synonyms, and the meanings of their synonyms, I would not have been able to write this book. I believe you will notice this fact now that I've pointed it out to you.

Had the process of writing this book been different than it was, the book itself would without a doubt be different, too. It likely wouldn't even exist.

To get the best results from reading this book, I recommend that you read it as I believe it was intended to be read—in one sitting.

Do that, and I believe the person who gets up from that chair will not be the same person who sat down in it. And I believe the life you enter upon rising will not be the same life you exited when you started reading. Some books have the very real power to change lives. I believe this is one of those books.

1
To Manifest and Create

To *manifest* something, as the term is often used to describe a metaphysical process, is to bring something forth, to make something available and present where it was not before available or present.

For our purposes here, to manifest something is to bring that thing into the realm of someone's material life where it can be experienced by that person and others. Some synonyms for the word *manifest* are *demonstrate, show, exhibit, reveal,* and *display*.

In metaphysical teachings, the word *demonstrate* has long been used to describe this process of manifesting as we are defining it here. And to *demonstrate* a thing can be defined as exhibiting it or making it evident.

Notice the fact that when we consider the meanings of the words *manifest* and *demonstrate*, what we are actually talking about is a process by which something that is not present or evident at the moment becomes present or evident. (With further research, you will discover the same basic shades of meaning for the words *show, exhibit, reveal,* and *display*.)

By definition, what we have been discussing here thus far seems to be quite a different process from that of creating something. When we *create* something, we bring that thing into being; we *cause* it; we *produce* it.

But to *produce* something can mean to either bring that thing into existence or to bring that thing forward and to make it available so that it may then be known, seen, comprehended, experienced, used, and so on.

So, the act of manifesting something can be said to be the act of making something present and/or the act of creating something and bringing it into being.

Now, what we are really talking about here is the act of creating something, of bringing something into being, because even if something that already exists is simply *made present*, for practical purposes, the resulting *circumstances* will have been created or brought into being where they did not exist before.

In the case of something that does *not* already "exist," we could ask this basic question in two ways: *At what point can a given thing be rightly considered to have been created? At what point can a thing be rightly considered to have been brought into being?*

Take the example of an oil painting in progress. At what point can that painting be considered to have been created? At what point can that painting be considered to have been brought into being? Another way to ask the question would be this: *At what point can that painting be said to exist as a painting?*

Would it be only after the last brushstroke has been made? If that were the case, would it mean the painting did not exist before the very last brushstroke was completed? Would it mean that until the last brushstroke the painting had not yet been created?

Would we be able to see the painting right before the last brushstroke? Yes, we would be able to see it. That would be proof to us that a painting had, in fact, been created, that a painting had been brought into being, some time before the last brushstroke was made.

But how long before the last brushstroke did the painting actually become a painting? Upon which brushstroke, out of the hundreds or thousands made, could we say a painting, as such, was brought into being or that a painting had, in fact, been created?

Some might say the "painting" was created at the point where what was in the process of being depicted on the canvas became self-evident. But if it was, for instance, an abstract painting, it would likely *never* be evident what the painting was supposed to depict.

So, how many brushstrokes would it take before we could say a painting has been brought into being or that a painting has been created? Fifty brushstrokes? A hundred? Two hundred? Would it depend on the painting? Would it depend on who you asked?

Look at the definition of *painting*, and you will find that, basically, a painting is a painted picture. Look at the definition of *picture*, and you'll find one definition to be that a picture is a painting, especially a work of art. Look up the word *art,* and you'll find art has to do with human creative skill or the application of that skill—or with a work exhibiting these qualities.

Let us now suppose an artist dips a paintbrush into black paint and with one flick of his wrist splashes paint across a blank canvas. Can the result be called *art*? And who's to judge what art is or is not? Artists have in fact splashed paint onto canvas and then sold those "paintings," those works of "art."

2
Brushstrokes

But let us get back to the question of *when*, exactly, a painting can rightly and accurately be said to have been created and to have come into being as a painting. On which brushstroke would that happen?

How about on the tenth brushstroke? How about the fifth brushstroke? How about after just one brushstroke? Say an artist, using black paint, makes just one brushstroke on a blank canvas. Can we then call that a *painting*? Would we say a painting exists?

Remember, it's just an otherwise blank canvas with one brushstroke on it. It might help us to agree that a painting exists if we could first agree that what we are looking at could be called *art*. But what would make a single brushstroke on an otherwise blank canvas *art*?

Would it depend on how long or wide the stroke was? Would it depend on the texture or the color of the paint? Would it depend on where the brushstroke was placed on the canvas? What prerequisites would have to be met before we would agree that even just one single brushstroke on an otherwise blank canvas could rightly be called art?

Remember, a *work of art* can be defined as a work exhibiting human creative skill. But is there any creative skill involved in making a single brushstroke on a blank canvas? If you look at the word *creative*, you will discover that it, in part, means *able to create*.

Consider a single brushstroke on an otherwise blank canvas. Was the artist able to create it? Yes. Has it now been created? Yes. Does it now exist? Yes. Did it require creative ability to create? By definition, yes, it did. Can it be called *art*? By definition, yes, it can be called art. But can it be called a *work* of art?

A *work* (noun) is something made by *work* (verb). *Work* as a verb means to engage in bodily or mental activity. Did the artist engage in bodily or mental activity to put just one brushstroke onto canvas? Yes.

Let us assume a painting will require a thousand brushstrokes to be completed. Would we say the first brushstroke on the canvas could be considered a work of art? Some people would likely think that it should not be called a work of art. But even if we *don't* call a single brushstroke on an otherwise blank canvas a work of art, would that necessarily mean, in and of itself, that a painting does not yet exist?

Now, let's say the artist dips a two-inch-wide paintbrush half in black paint and half in yellow paint and makes a six-inch-long brushstroke at a forty-five-degree angle four inches left of center on an otherwise blank canvas. He then steps back and, looking at the canvas with a satisfied expression, blurts out, "It's *done!*" By definition, that single brushstroke on the canvas would, in fact, be a "work of art."

So, at least we have now established that a single brushstroke on an otherwise blank canvas can, in fact, be rightly called a work of art, and so, by definition, this work of art could also be rightly called a painting.

But back to the first brushstroke of a thousand-brushstroke painting. When can *that* work be said to exist as a work of art or even just as a painting?

Imagine you are looking at a half-completed painting of a landscape. Imagine you can clearly see the beginnings of mountains in the background, a river in the foreground, and so on. Do you think you'd say that a painting does not exist? Probably not. You'd likely say a painting exists but has not yet been completed.

For that matter, do you think you would be inclined to say a work of art does not exist? If we can agree that a painting exists, we can likely also agree that a work of art exists, even though that work of art is still in the process of being created and has not yet been completed. But how long before that would we be able to agree that a painting exists? At what point will we be able to agree that a painting exists as a painting?

It's not unreasonable to conclude that even a single brushstroke of a thousand-brushstroke painting would be clear evidence that a painting (and even a work of art) *does exist*, though it is still admittedly in the process of coming into being and of being created.

The painting *exists* because the painting doesn't *not* exist. A painting has been begun; therefore, a painting can now be said to have come into existence.

3
No Brushstrokes?

It would be easy to conclude from all of this that a painting comes into existence upon the completion of the first brushstroke. The unstated assumption would be that *until* the first brushstroke is completed, no painting exists. At first glance, that assumption would indeed seem to be logical and make sense. But wait...

Now, what if the artist was interrupted in the middle of that first brushstroke and walked away without completing that stroke? What if he only got to complete half of the intended brushstroke? Could we credibly say a painting exists if there were only one half of a brushstroke on an otherwise blank canvas?

Going by the logic we have already followed up to this point and by the conclusions we have already reached, it seems we would indeed have to agree that even *half* of a brushstroke on an otherwise blank canvas would be evidence of the existence of a painting.

And we would have to agree that even a *quarter* of a brushstroke would be evidence of the existence of a painting. We would agree that even a *tenth* of a brushstroke would be proof that a painting—although undeniably unfinished—existed.

So, we might conclude from all of this that a thousand-brushstroke painting comes into being upon even the smallest percentage of a brushstroke being made on a blank canvas. The unstated assumption would be that no painting exists *until* at least the smallest percentage of a brushstroke has been made. At first glance, that assumption would indeed seem to be logical and make sense. But wait...

Let's say the artist is standing in front of his blank canvas holding a paintbrush that he has just dipped into paint. He is about to make his first brushstroke. His arm is bent at the elbow, his hand in front of his chest. He takes a deep breath. Then he turns his head and upper body slightly in the direction of a window and glances out at the peaceful setting of his yard.

As the artist's upper body and the hand holding the paintbrush swivel slightly toward the window, one single bristle, bent and sticking out from the rest, ever so lightly makes contact with the canvas, leaving only one tiny, barely perceptible speck of paint. There is just one tiny, barely perceptible speck of paint, and no part of a brushstroke has yet been made. Would we say *that* shows evidence of the existence of a painting?

Going by the logic we have already followed and by the conclusions we have already reached, it seems we would indeed have to agree that without even the tiniest percentage of a brushstroke being made, and with only the tiniest speck of paint having been transferred from the brush to the canvas—and even though perhaps inadvertently—a painting has come into being. We could agree that a painting now exists.

So, now we are at the point where we have agreed that a thousand-brushstroke painting can come into being even with no brushstroke ever being begun or attempted, as long as at least the tiniest speck of paint has been transferred from the brush to the canvas.

As you may have guessed, the unstated assumption would seem to be that no painting exists *until* at least *some* speck of paint has been transferred from the brush to the canvas. At first glance, that assumption would indeed seem to be logical and make sense.

Now, let's really think about this thing for a moment. While the canvas is completely blank and the brush is free of paint and the paint is still on the palette, or even still in the *tubes* for that matter, could there be any way we could credibly say a painting exists?

It would seem not. After all, there would be no evidence of a painting or of a painting in progress. If a painting is a painted picture, and no paint has yet touched the canvas, one might naturally conclude that there is not yet a painting. One might confidently conclude that a painting does not exist.

It doesn't seem to make sense that a thousand-brushstroke painting could exist before even a tiny speck of paint has been transferred from a brush to a canvas. This line of reasoning and discovery may have come to its seemingly inevitable limit. This mental journey could end right here. But wait...

4
The Foundation

When we look up the definition of the word *exist*, we find it means to have a place as part of objective reality. And when we look up the word *objective*, we find that if something is objective, it is *external to the mind* and *actually existing* and *real*.

So, would we say the "painting" in question does *not* exist since it has no place in objective reality because *objective* is supposed to mean *external to the mind*, *actually existing*, and *real*? It might appear that way. But let's look at this a bit closer. We don't have to be strictly bound by definitions that may not hold up under all circumstances or closer inspection.

We've looked at the *objective* in *objective reality*. Now let's look at the *reality* in *objective reality*. *Reality* is defined as *what is real* or *what underlies appearances*. But with no paint yet on the canvas, it would seem there is not even the slightest *appearance* of a painting. Look up the word *appearance*, and you will find it means, in part, an act of or an instance of appearing. By definition, for something to *appear*, it has to become visible or be visible. Consider the fact that something fits the definition of being *visible* if it can be *seen* or even simply *perceived*.

According to one definition of *perceive*, something can be perceived by being apprehended either through the sight or with the *mind*. Look up *apprehend*, and it sends you straight back to *perceive*.

So, by definition, if something can be perceived by at least the mind, if something can be apprehended by at least the mind, it exists in *reality*. If something can be perceived, then it is in fact some *thing*. In order to be a thing, a thing must exist. And if something exists, it is part of reality. It is real and it can be perceived.

So now we understand that something needs only to be perceivable by the mind to be a part of reality. *Reality* is that which is *real*, *existent*, or *underlies appearances*. By definition, something *underlies* something else by being the *basis* of that thing or existing beneath the superficial aspect of that thing.

By definition, the *basis* of a thing is the *foundation* or *support* of that thing. And what, exactly, is the foundation and support of the painting before the first speck of paint has ever touched the canvas? It's not the canvas. It's not the paintbrush. It's not the paint. The foundation of the painting before the first speck of paint has ever touched the canvas is the *thought* of the painting. The foundation of the painting is the *image* of the painting lying within the artist's *mind*.

The image of the painting can be perceived, so by definition, it is *visible*. It is true that it is not visible to the physical sight, but it is visible just the same. It is undeniably *visible*. It can be *seen*. It can be *perceived*. It exists in *reality*. So, it *underlies* the *appearance* of the painting—even before the painting has appeared.

The painting *exists* before one speck of paint ever touches the canvas. The painting is *real* before one speck of paint ever touches the canvas. The painting *exists in reality* before one speck of paint ever touches the canvas. The painting *is* before it ever *is*.

Things *are* before they *are*. For something to be able to come into material existence, it must already be *something somewhere*. The painting that dwells within the mind of the artist is actually as real as a completed landscape painting done in oil paint.

This is not just some fanciful theoretical conception or elaborate intellectual concoction. And these conclusions are not just the manufactured products of the creative interpretation of the officially sanctioned definitions of words—as powerful a case as that would make in and of itself. That's not nearly the extent of it.

No, it goes deeper. The mystics will nod their heads in agreement at this point. After all, they've been telling us this sort of thing for as long as they've been around.

And even the scientists can lend us a hand here. They can explain that matter and energy are the same substance, only in different states of being. They can explain that matter and energy are interchangeable and that matter and energy do, in fact, interchange.

5
It Already Is

With each passing year, with each passing day even, the wall separating the mystics from the scientists is dissolving just that much more. Matter and energy are interchangeable. Matter and energy are, in essence and at their very core, the *same thing*.

The *thought* of the landscape in oil is made up of the *same stuff* that makes up the oil painting itself. The thought of the painting and the painting itself are in different states of being, but they are both comprised of the same substance, and they both exist in reality.

This book is called *Advanced Manifesting Made Easy*. One reason manifesting can be made easy is that the thing one wants to manifest *is already real*. Think on that a bit. Let that sink in. As an example, that life as a multimillionaire you have fantasized about living *already exists*. You are *already living it*.

Know this: *The moment you conceive of a thing, that thing exists.* A case could be made that a thing *already* existed even before you consciously conceived of it. But there is no need to go any further along that line of thought than we have already gone. It would not serve sufficient practical purpose as to be worth the time and energy it would take, and it would be beside the point for now.

Thus, and again, if something you would like to manifest for yourself—let's assume a six-bedroom house this time as another example—already exists for *you*, at least by the time you have conceived of the concept, then it is *already yours*. You *already have it*.

Some genuinely sincere and dedicated students of metaphysics, once having been instructed to *believe* that they *already have* the thing they want to have, sometimes think to themselves or say something along this line: *How can I really believe I already have a thing I know I don't actually already have?*

So many of these students still do not understand that they *do* in fact already have the things they want. There is nothing to *believe*. It already *is*. It's not something to be believed—it's something to be *known*.

Application of mind-power techniques in harmony with metaphysical principles and universal law for the purpose of manifesting desired outcomes is not about *creating* something as much as it is about letting something that already exists appear to your physical senses in the material dimension. Matter and energy, energy and matter—they are the same thing.

Thoughts are things. Things were once thoughts. Things *are* thoughts *objectified*. Thoughts can exist independently of their material-world counterparts. But material world things and circumstances cannot exist independently of the thoughts that give rise to them, as the thoughts are the *basis* of, the *foundation* of, the things and the circumstances they *underlie*.

So, if we accept that as being true, it would mean that the thoughts behind the things can be said to be more real than the things themselves. Think about *that*.

You are presumably reading this book because you are interested in the idea of your manifesting being made easy. If something is *easy*, by definition, it is not difficult and can be achieved without great effort.

If you want your manifesting to be not difficult and to be achieved without great effort, come to know that what you want to have already exists. Don't be fooled by the illusion of time; those things you seek are not in the future. Don't be fooled by the illusion of space; those things you seek are not at a distance from you.

The things and circumstances you want are *here*, and they are here *now*. A pivotal action you can take in this moment is to release the false idea that you do not yet have the things you want to have. Let it go. Be free of it. Stop *trying* to make things happen—allow things to be what they already are. Stop *trying* to make things appear—see that the things you want are already here.

Think. Do you want your manifesting to be easy? Then stop making it hard. Remember, if something is easy, it is by definition not difficult and can be achieved without great effort. Do you want your manifesting to be easy? Then stop *trying* to do it. Just do it. Or, better yet, just *allow* it to *happen*.

6
Oneness

How hard do you have to try to be able to keep your eyebrows on your face? Say, on a scale from one to ten, how hard is it? It would be zero on a scale from one to ten. Wouldn't it? You don't have to *try* to keep your eyebrows on at all. They're already on. They're a part of you. They are not separate from you.

How hard would you have to try to get, as an example, the new car you have been wanting? That would be quite a bit harder than trying to keep your eyebrows on. Wouldn't it? After all, the new car is *separate* from you. Isn't it? We shall see.

Would you like your manifesting to be made easy? Peer beyond the veil of the illusion of separateness. Begin to peer into the glorious truth of Oneness. It's a concept that has been around for countless generations. Throughout human history, only a select few have really understood the reality of Oneness. To the average person, Oneness is about as foreign a concept as you can find. After all, there are billions of people on this planet. Not just one. And there are countless objects in existence. Not just one. And there are endless locations out there. Not just one.

Think about it. One of the most basic assumptions that goes into the makeup of the average person's sense of reality and sense of self is that they are separate from all they perceive as being outside of themselves. They think along such lines as these: *I am me and they are them. This is me and that is not me.*

People do not as a rule challenge this assumption. People do not as a rule even *think* of the *option* of challenging this assumption. And really, why should they? The obvious is so obvious it is beyond question.

But *is* the obvious really so obvious as to be beyond question? If you have read much from metaphysical and spiritual writings, you know there has been another theory circulating out there. You know the theory has been presented—time and time again and over countless generations—that separateness is an illusion and that Oneness is the actual, valid reality.

So, what if this theory is true? Some people, upon hearing this theory of Oneness, take to it right away. They intuitively feel it to be true. They can sense the oneness of creation. They can sense *their* oneness with all that is. And then there are those people who are hesitant at first but eventually come around and do end up accepting and even embracing the basic premise that there is only Oneness and that all of animate and inanimate creation, all of seen and unseen reality, and all of formed and unformed substance is part of that grand, overriding Oneness.

And then there are those people who just plain don't like this idea of Oneness. It upsets them to even hear about it or think about it. Many will fiercely defend their belief that they are separate from all they see.

It's a good thing a person can obtain measurable results using metaphysical techniques for attraction and manifestation *without* having to subscribe to the theory of Oneness. A person can use mind power and the techniques of practical metaphysics, and they can get many of the things they want to get; they can bring into being circumstances they want to create.

Some people attempt to align their actions and habits with the natural workings of the unseen forces of the universe, and depending on how well they do that, they get their desired results. They manifest.

But is there another way to manifest? Is there a more *advanced* way to manifest? Is there an *easier* way to manifest? Yes, there is. It's not for everyone, but there is a more advanced way to manifest. There is an easier way to manifest. Part of that way is to realize your oneness with all you wish to acquire or bring about and then to *allow* what is already yours, to *allow* what is already *you*, to just naturally take form in your life.

More than a function of belief, it's a function of desire and will and intention. Think. You don't *believe* you can scratch your nose as much as you *know* you can scratch your nose. Right? And any time you have the intention of scratching your nose—which always has behind it the *will* to do so, which always has behind it the *desire* to do so—you simply scratch your nose.

7
Scratching an Itch

Belief is clearly a powerful spiritual force. Using belief, one can cause things to happen—one can indeed *manifest*. But to bring oneself to believe things are a certain way when they do not appear to be that way requires a certain amount of real effort.

Now, it will most often require much less effort to acquire a given thing through the power of belief than would be needed to acquire the same thing using money earned from days, weeks, months, or even years at some job or business but, still, it's effort.

The way most people tend to go after what they want is much more a process of getting than it is a process of receiving. It's much more a process of reaching out for something than it is a process of opening up and allowing something to come in. Two different things.

Belief is the acceptance of an assertion or idea as being true, regardless of whether or not that assertion or idea *is* true. But *knowing*, on the other hand, is a state of being aware or informed. Being *aware* is being conscious and having knowledge. And being *conscious* brings us right back to being awake and aware. So, we can only know things that are true. We can believe just about anything, and many people do.

Think about how natural it is for you to scratch your nose. You don't try to cultivate the belief that you can and will scratch your nose. You don't do affirmations or visualization in an attempt to bring about the scratching of your nose. The very idea of it is absurd.

The way it goes is that your nose itches, you desire to scratch it, you will to scratch it, you intend to scratch it, and then you scratch it. It may sound like a series of separate steps, but in practice, we know it to be one simple action accomplished in one smooth motion.

Thinking is involved but not as a conscious effort—the thinking is a natural, instinctive, nearly effortless response to the itch, which leads to the desire then the will then the intention then manifestation—the nose gets scratched. What takes place between the itch and the scratching of the nose is barely even noticeable.

That is how manifesting is made easy. *That* is how manifesting is made into a nearly effortless, spontaneous act instead of being a planned-out series of individual steps requiring deliberate and obvious effort to accomplish. Now, you do have to move your shoulder, your arm, your hand, and perhaps one or two of your fingers in order to scratch your nose. And that requires effort. But it's largely unconscious effort, not effort propelled by deliberate, conscious will. It's lack of effort propelled by barely conscious will.

You never have to think about it. Upon perceiving an itch on your nose, you don't think to yourself, *Oh, now this is just grand! My nose itches, and now I'm going to have to move my shoulder and my arm and my hand and perhaps even one or two of my fingers to scratch it. What a hassle!* No. It's nothing like that.

There is one seemingly seamless string of events, most of which you are not even consciously aware of, or are only minimally aware of, in the moment—itch, desire, will, intention, scratching. The whole business is over before you know it, and you likely didn't even have to stop whatever other task you may have been doing at the time. For all practical purposes, you could say your nose scratched *itself*.

What other itches—besides the occasional one on your nose—do you have? Do you have an itch for a better job or a new career? Do you have an itch for a larger home or a new car? Do you have an itch for better health or more free time? Do you have an itch for an increased income or a greater net worth? Do you have an itch for the ideal mate? Do you have an itch to travel the world? Do you have an itch to help financially support certain people or organizations?

Think about why your nose is so easy to scratch. One reason is that it is, in fact, *your* nose. It's a part of you, and so you can do what you want with it. Another reason is that it is always within reach. And yet another reason your nose is so easy to scratch is that the act of scratching your nose takes place mostly of its own volition with little if any conscious effort.

8
Observation Equals Influence

What if you knew that the objects of your other itches are also, in fact, yours? What if you knew that the other things you itch for are also a part of you and that you can do what you want with *them*? What if you knew that those other things you itch for are always within reach? And what if you knew that the acts of scratching those other itches can take place mostly of their own volition with little if any conscious effort?

How would that make the process of attaining the things you desire different? Would it make that process easier? Does it sound like a more advanced and easier way to manifest?

Now, even with this more advanced and easier way of manifesting, sometimes certain things must be done. Plans must be made and carried out. Information must be gathered, assimilated, and acted upon. Phone calls must be made. Letters and emails must be written. Just as you must move your shoulder, arm, hand, and finger to scratch your nose, you often must perform certain acts to scratch your other itches.

But as in the example of scratching your nose, the effort required will be greatly minimized because you will be in the flow of the natural process, and so you will be motivated to take the necessary actions instead of having to force yourself to take those actions. Instead of having to push ahead, you will be pulled forward. Itch, desire, will, intention, and then manifestation—your itches get scratched.

And even when you are using this advanced and easy way of manifesting, you will still be free to use mind-power and metaphysical techniques as they relate to the processes of attraction and manifestation. But such actions will tend to rise up from within you as a natural part of the experience of being what you are, as opposed to you having to put forth conscious will and effort to try to *do* something.

You will not use the various metaphysical techniques as a person who is trying to be, do, or have certain things. Instead, you will use these techniques because it is how a person who is already being, doing, and having those things would just naturally function and behave. It may sound like a subtle difference, but it is a *big* difference just the same. So, let's move on.

You see things that exist. You see things that happen. Things are and you observe them. Events transpire and you observe them. Things are what they are, and you are able to perceive them. Some of us can accept and even embrace the notion that we are one with all that is—that we are a single cell in the universal mind-body. But can we take this core concept any further?

Modern science has shaken the very foundations of what we human beings have for so long accepted as our material reality. Quantum physicists, as a result of their experiments, have called into serious question the validity of our familiar and long-held worldview.

One unquestioned and mostly unspoken aspect of the commonly held world view is that things are and things happen and we can observe what is and what is happening, but we do not affect what is and what is happening simply by observing it—we merely perceive and apprehend what we become aware of.

That has been another one of those basic assumptions we have had that rarely, if ever, is challenged by the average person. But what if this assumption is wrong?

We've already established, at least in theory, that we are One, that the individual is, in fact, an actual part of all other people, things, and places. So, we might say that, still, we *appear* to be separate, and we can look out at the other parts of ourselves and see what is going on out there and not—without direct action—have a material effect on what is going on out there.

That sounds plausible. Logical even. The findings of quantum physics, though, lead to the apparently unavoidable conclusion that we cannot *merely* observe *anything*. Quantum experiments have shown that an observer (a quantum physicist, for example) cannot engage in the act of observation (of an experiment, for instance) without materially affecting that which is being observed, without *influencing*—at least to *some degree*—that which is being observed.

9
A Great Responsibility

Through experiments in quantum physics, we have come to the seemingly unavoidable conclusion that the act of observation is a *creative act*. In other words, through observation, we help to create, recreate, perpetuate, increase, lessen, and end that which we perceive and observe. Observation equals influence. Observation equals creation.

This means that we have not been merely *discovering* what is so through our observations of it—we have been helping to *make* what is so *what it is* through our observations of it. Let that concept sink in. Let it settle. Begin to *feel your power*.

So, in which ways do we affect those things and circumstances we observe? We affect them in an infinite number of ways. And by what means do we affect those things and circumstances we observe? Our power is in our thoughts, feelings, beliefs, expectations, opinions, attitudes, and assumptions; it's in our prejudices, desires, fears, intentions, understanding, and confusion; it's in our *knowing*.

We affect all we perceive in countless ways. By how we observe, we affect people, objects, events, and circumstances. We make things more desirable or less desirable, larger or smaller, stronger or weaker, better or worse. We make things stop or continue on.

For most, of course, this ongoing process of modifying all they observe is unconscious. To some, this process is one they know is going on, but they have no concept of *how*—in which specific *ways*—the act of their observation is affecting what they are observing, and they have no conscious *control* over the process.

We can all hope our observations will have a constructive—and not a *de*structive—effect on that which we observe. But hope is not knowing; hope is not doing.

One question we could ask is this: *How is my observing affecting what is going on out there?* Just as important a question is this: *How is how I am affecting what is going on out there affecting* me?

And it *is* affecting you. How could it *not* affect you? You are one with all you perceive and even all you do not perceive. You affect everything that is, and everything that is affects you. One moment of your uncontrolled anger will reverberate out to the ends of all creation; one moment of your unconditional love will also reverberate out to the ends of all creation.

Anything that is helpful to you is helpful to everyone and everything. Anything that is harmful to you is harmful to everyone and everything. Anything that is helpful to anyone or anything is helpful to you. Anything that is harmful to anyone or anything is harmful to you. Ponder this deeply.

By learning how to manifest in the advanced and easy way, you will not only be helping *yourself* in countless, meaningful ways, but you will be helping all people on the planet in countless, meaningful ways.

In light of this encouraging fact, consciously and intelligently manifesting in the advanced and easy way should *not* be viewed as a selfish act—instead, it could be viewed as a humanitarian effort.

Improve your life to whatever degree, and to that degree you improve the whole world. Increase your happiness to whatever degree, and to that degree you increase the happiness of the whole world.

Regarding people who understand this, the argument can be made that not only is it their great blessing to be able to improve their lives and themselves by manifesting in the advanced and easy way, but that it is also their great responsibility to do so. A single individual has the potential to save the world; a single individual has the potential to destroy the world.

We can help the world to be happy not so much by fighting things that make people *unhappy* as we can by becoming happy *ourselves*. We can help the world to live more abundantly not so much by fighting things that breed lack as we can by living abundantly *ourselves*. And, of course, by sharing our abundance in whatever forms it may come. Know that what we *resist persists* and that we give *might* to what we *fight*.

10
Point of View

As individual conscious creators of material reality at large and of our own personal material reality, we should strive to move toward what we want more than we move away from what we don't want. We should strive to increase what we view as desirable things for the world and for us more than we strive to decrease what we view as undesirable things for the world and for us. We should be for what we deem as good, right, and just more than we are against what we deem as bad, wrong, and unjust.

This does not mean we cannot and should not *take actions* that may make us appear to be fighting or resisting something. It means we should take all such actions from a fundamentally different philosophical point of view.

For example, it's OK for you to give money to homeless people. But you should not give homeless people money in an attempt to compensate for their lack. Instead, you should give homeless people money with the clear intention of adding to their abundance.

Don't see that a certain homeless person has little or no money. Instead, see that the homeless person in question is in the process of receiving all the money they need, and see the act of you giving that person money as evidence of that fact. In other words, don't feed people's poverty. Instead, feed their prosperity.

Yes, you can give a homeless person money, and it is an admirable thing to do, but there is a harmful way and a helpful way to do it. In the one way, you observe the situation in a manner that helps to push the homeless person further down into their prison of lack and despair. In the other way, you observe the situation in a manner that helps to lift the homeless person up into the freedom of abundance and hope.

Though the act itself is important, of greatest importance is the consciousness behind the act. Thought and feeling are the forces that create and recreate material reality—for good or bad.

Always remember that it is how you *view* what you observe that determines what your effect will *be* on what you observe. Observing is creating.

Think about how you normally look at things such as war and violence, illness and disease, and so-called injustice. How have you been viewing such things? How have you been observing them?

Have you been observing them in ways that help strengthen and perpetuate them or in ways that help weaken and lessen them? It's one or the other.

The gift of the power to observe carries with it the great obligation to observe responsibly. The gift of our power to transform our lives by how we observe them is our gift of the power to transform the *world* by how we observe *it*. And just what is it at the bottom of all of this that determines *how* we observe a given thing?

In the final analysis, it comes down to the basic factors of individual will and personal choice. We can choose how we view things, and by how we view things, we will make things what they *are* and what they *will be*.

Can you grasp the significance of what you are reading here? Do you have any idea at all of how powerful you are? Do you know that your life will be what you choose it to be? Do you understand that you have thus far realized only the tiniest fraction of your potential? Can you fathom how much more is possible for you?

Do you *see* that what you *see* will be what it is in accordance with *how you see it*? Can you accept the fact that you can see things in any light you decide to?

This book is not so much about teaching you what to *do* so that you will be able to manifest in the advanced and easy way as much as it is about reminding you of what you *are* so that you can manifest in the advanced and easy way as a natural *result* of what you are.

It's time to wake up. You need to know yourself. You need to stop hiding from yourself. You need to stop denying yourself. You need to allow yourself to *be*.

11
Choose

You are, presumably, reading this book to learn how to manifest in the advanced and easy way. Perhaps you want more money or more success or more influence. Maybe you desire more fulfillment in your life or better relationships or greater peace of mind. The point is there are things you want to attain that you think will give you other things you are seeking.

For instance, perhaps you think that finding the love of your life will bring you happiness or that establishing yourself in a lucrative profession or amassing a great fortune will give you security.

On the *surface*, perhaps, you may be able to find some of what you are seeking in this way—at least in the short term. But you will never truly gain inherently inner qualities—such as the emotion of happiness and the feeling of security—from temporary outer things such as relationships, objects, and circumstances.

Do you want to be happy? Then be happy right now. Do you want to feel secure? Then feel secure right now. Have you done it yet? Are you happy and secure now? All that stands in your way is a decision.

You may think you can never be happy until you find your ideal mate. You may think you will never feel secure until certain financial conditions have been brought about. But what if that's not how it works? What if you first need to be happy before you can find your ideal mate? What if you must first feel secure before you can bring about certain financial conditions you desire? What if you have it backwards?

Entertain the following concepts. The perfect mate may not be the *cause* of your happiness but an *effect* of your happiness. Having the financial resources you desire may not be the *cause* of your feelings of security but an *effect* of your feelings of security.

The universe doesn't bring you what you want or what you need as much as it brings you what you *are* and what you are *made of*. Everything you become aware of and experience is *you* being reflected back to yourself. All you can ever see and know is *you*.

Do you want to start manifesting in the advanced and easy way? Then you must stop doing it the old, hard way. Do you want to start manifesting in the advanced and easy way? Then you must cease *being* in ways that keep you from doing so. Do you want to start manifesting in the advanced and easy way? Then *choose* to manifest in the advanced and easy way.

Do you want things to be easy, or do you want things to be hard? Do you want to *have* the good things you long to have, or do you want to continue to *not* have the good things you long to have? You must decide what you want your circumstances to be.

Do you want to live the life you *want* to live, or do you want to live a life you *don't* want to live? The choice is yours. In *this moment*, you can decide, and in *this moment*, you can make the fundamental mental and emotional adjustment that will transform your life.

If you have decided to have what you want and to manifest it in the advanced and easy way, then congratulations. This can be the defining moment of your life. This can be the moment in which everything changes. This can be the moment in which you go from *not* being, doing, and having what you want to *being*, *doing*, and *having* what you want.

Your life is about to change, and it will never be the same—*if* you *decide* to act in this moment, and *if* you make this one fundamental change in how you see yourself: Choose to see yourself as a person who *is* all you desire to be, do, and have. That's it. There's nothing to get. There's nothing to do. Just *be*.

Again, choose to see yourself as a person who *is* all you desire to be, do, and have. Take a look around you. Do you like what you see? Is your home an accurate reflection of who you are or who you would like to be? Are you truly happy and content in it? Are your relationships as you would choose them to be? Is your job or profession what you really want to be doing?

12
In This Moment

Yes, take a good look around you. Things may be about to change. If you will take this opportunity to choose to see yourself as a person who is all you desire to be, do, and have, then your circumstances must now begin to transform into a more accurate reflection of the present reality of what you know yourself to be—the present reality underlying the appearances that have not quite yet appeared. A new foundation will give rise to a new structure.

If you've made the switch in perspective, you may be feeling pretty good about yourself and about your life right now. You may be feeling your power at a deep, visceral level and in a way you have never done before.

In this moment, you may have arrived at exactly where you need to be so you will finally be one with yourself, and so, one with the universe, and so, one with all you wish to bring into your experience.

Congratulations. You are now a person who is manifesting in the advanced and easy way. In this moment, you have become all you can be in this moment. This is cause for celebration. So celebrate. Feel it. Breathe it in. What a glorious moment this is.

Yes, what a glorious moment that was. But that was then and this is now. You presently exist within the illusion of time. Moments come and moments pass. They come and they pass. They come again and pass again. If you did as you were instructed, you were in that moment all you could ever be in that moment.

But what about *this* moment? What are you in *this* moment? Are you still all you could ever be? Or are you less? Has doubt started to creep in? It will, you know, start to creep in. Are your old, habitual negative thought patterns struggling to reassert themselves? They will, you know, struggle to reassert themselves. At least for a time.

You might have thoughts such as these: *Who do I think I am? What do I think I am? This time I've gone too far! This is all just wishful thinking!* Do you really think you could be all you've now been told you can be? For your sake, you should hope and pray you do.

Moments come and moments pass. They keep coming and they keep passing. Or do they? A person might conclude from all of this that one way to proceed from here would be to, in every moment, remind themselves of who and what they are. That would indeed be better than allowing themselves to forget.

But another and even better way to proceed is to step outside of time. *Will* your very essence to step off the merry-go-round of the endless succession of moments and into this, the only true moment there has ever been—this glorious, timeless moment.

On the merry-go-round of time, you go around and around and you feel like you're going somewhere, but you never *do* go anywhere. Off the merry-go-round of time, from the sidelines of time, you see everyone else going around and around and getting absolutely nowhere. There's nowhere for them to go. Each person is already in the only place they could ever be.

From your vantage point outside of time, the main thing you must continue to do is to simply cling to the present, timeless moment. Clinging to the present, timeless moment will require you to maintain your focus in a way you may not be accustomed to doing. That will be a challenging task in the beginning, but the more you do it, the more habitual this state of being will become.

Now follow this. If, a few paragraphs back, you experienced a moment in which you were all you could ever be, and you are now outside that moment, you will never be able to get *back* to it. And you will not be able to gain that insight in some subsequent moment. There was only that moment.

But in reality, that moment will never cease to be. And you are still, in fact, *in* that moment. *That* is what is real. *That* is all that could ever *be* real. Everything else is *unreal.* Everything else has always *been* unreal.

13
The Reflection

What is real? You are real. You are everything in its entirety, though for whatever reason the *you* that you know yourself to be is presently peeking out at itself from this far corner of itself. What is real? This moment is real. This moment is the only moment that has ever been and the only moment that can ever be.

Everything seemingly outside of you is *not* outside of you. Everything seemingly outside of this moment is *not* outside of this moment. There is only you and only this moment. This is you and this is your time, and you are and will be what you decide you are and will be. Only you can decide, and you can only do it now.

Before now, you would likely have assumed there was no painting on any blank canvas you would have happened to see. But as you now know, there *are* paintings on certain blank canvases. Those paintings are in the unseen and have not yet come into the seen in relationship to linear time, but the appearances of things in so-called reality are not actually things at all—they are only *reflections* of things that *are* things.

The real painting lies behind the appearance painting and is within the energy realm of the artist's mind and, indeed, within the mind of the universe. The real painting may or may not be consciously fully formed yet, but it is there just the same. The real painting may not even be consciously recognized or acknowledged by the artist yet, but it is there just the same.

The material painting will appear to physical sight in due time, but it will not be the real painting at all—it will be only a mirror reflection of the real painting. The material painting could be destroyed, and it could then be re-reflected back into this dimension from the mind and through the hand of the artist.

Would that painting be an exact duplicate of the first material painting in every detail? No, it would not be. Even if it would appear to be so to the sight, it would not be. But the original material painting was not an exact duplicate of the underlying energy painting.

Understand that Spirit and matter are written in two different languages, and it is inevitable that something will be lost in the translation from one language to the other. Yes, a material painting is a reflection of an energy painting. It's a mirror image.

But mirrors have imperfections. Mirrors get dirty. Mirrors fog up. Don't be fooled by appearances. Don't mistake the reflection of the thing for the thing itself. Don't be fooled by the appearances of your life circumstances. Don't mistake them for your life itself.

Stay in this timeless moment, know what you are, and understand that if what you see is less than what you want to see and have the intention of seeing, it's only because the reflection before you has not caught up with the new reality underlying what your life now is.

In the material dimension, which is clouded by the illusions of time and space and by the limits of human perception and understanding, there is routinely a time delay between what is and what appears. Things from beyond time and space do come to appear within time and space, but it's a process—be at peace with it.

Yes, be at peace with the ongoing process of material creation and manifestation. Know that things *are* before they *are*, time being a key element separating the two states of being. So, again, be at peace with the process. But never *forget* what the process *is*.

Don't allow yourself to be fooled by appearances. What you are now seeing before you are but the fading reflections of how you once thought things were. What you will now begin to see coming into focus before you will be the reflection of the new, present reality of what you have allowed yourself to become.

And what will this new reflection be? You are the only person who can answer that question. When you know the answer to that question, it must become so.

The painting of your life you now see before you is *changing* because you the artist are changing. And the deeper truth is that the painting has already changed because the artist has already changed. You the artist are, in this moment, all you can be. In due time, the reflection will show that reality more fully.

The reflection appearing in this dimension will never be a perfect duplicate of the core reality underlying it. The reflection will never be fully formed. Yet, the true reality it reflects will be revealed more and more.

14
How You See It

If you were somehow able to reflect all you are into this dimension, this dimension could not stand the shock; this dimension could not contain your Light.

This is indeed a process. You are becoming more and more of what you already are and what you are beginning to know yourself to be, and that reality will continue to be reflected into the world of appearances.

Do you have an itch? Do you have an itch for fame and fortune? Do you have an itch for material affluence? Do you have an itch to be of service? Do you have an itch to have more free time? Do you have an itch for spiritual growth? Do you have an itch to have more influence? An itch for love? Do you have an itch? Yes, of course you have an itch. You have *many* itches.

Don't try to scratch your itches. Instead, allow your itches to be scratched. Itch, desire, will, intention, and manifestation. It's just one seamless motion. It's just one smooth, nearly effortless action taking place through and within the medium of time and space.

Again, it's a process. And you, a single individualized cell in the limitless mind and body of the overriding One, are at the center of this process. And although you are contained within One, One is also contained within you. Again, you are at the very center of the process of creation. Can you feel your power?

Everything that has come before has brought you to here and to now, the place you know you've always been and will always be. Do you want to live the life you desire and to manifest it in the advanced and easy way? Then live the life you desire and manifest it in the advanced and easy way. You know what you are—now *be* it. You know what you want—now *be that, too.*

And always remember that you, the observer, are not separate from that which you observe and that you, the observer, are continually molding all you observe according to *how* you observe it. It really is all in how you see it. And you can see it how you *choose* to see it.

The time has come for you to allow your reality to be revealed to you. The time has come for you to allow your reality to be reflected out into the world of appearances so it may be revealed to one and all.

The time has come for you to unveil your glorious painting. Pull back the dark, heavy curtain that has been hiding the amazing Being you now know yourself to be. Let your eternal light shine out into the world in this, the only moment you will ever have.

Again, everything that has come before has brought you to here and to now, the place you know you've always been and will always be. Do you want to live the life of your dreams and to manifest it in the advanced and easy way? Then do it. Or, better yet, just allow it to happen. Can you feel your power?

AFTERWORD

Thank you for reading this book. I hope you have enjoyed it. And I hope you will continue to benefit from having read it. I have benefited greatly from having read countless books over the years.

I began to find my first self-help, spiritual, and metaphysical books in my early twenties, not long after I moved from New Jersey to California to try to find my way in the world. Before that move, I had no idea such books even existed.

And honestly, were it not for such books and my intense desire to learn, to grow, and to improve myself and my circumstances, I would have gone down a completely different road in life—a road I would rather not even think about or imagine.

Who could deny the assertion that books can and do change lives? It is my mission to write some of those books that do indeed change lives. I want people's lives to be better because I lived and because I wrote.

There are reasons I came into this life, and writing is one of them. I am living the life I was meant to live, and it is my sincere desire that you will live the life you were meant to live.

Can I ask two favors of you? First, if you think this or any of my other books can help people in some of the

Afterword

ways they could use help, will you help spread the word about me and my writings? You could do that by loaning my books to others, giving my books as gifts, and by telling people about my books and about me. By doing these things, you will bless me beyond measure, and I truly believe you will bless others beyond measure as well.

Second, please consider writing an honest review for this book. Reviews are important to the success of any book and any author. And reviews really do help people decide whether or not a certain book is right for them. So, by writing a review, you will be helping me personally and other people as well.

And speaking of those other people: Say your review is the one that causes a person to actually buy this or any of my other books. And suppose that person then reads the book. And suppose that book helps that person to substantially improve their own life and the lives of others. Just imagine the possibilities—lives made better because you wrote a book review.

Currently, Amazon.com is the most important place you could post a review, but feel free to post a review anywhere you choose. And keep in mind, even just a sentence or two could be sufficient. The number of words in a review you write is less important than what those words say. If you do end up writing a review for this book or any of my other books, feel free to let me know by contacting me on one of my social

Afterword

media pages, by email, or however else you can. I will enjoy hearing from you and reading your review.

Finally, always remember, you are capable of so much more than you have ever imagined. Learn, believe, act, and persist. Do those four things, and nothing will stop you from continuing to build a better and better life for yourself and for those you care about.

Peace & Plenty . . .

ABOUT THE AUTHOR

James Goi Jr., aka The Attract Money Guru™, is the bestselling author of the internationally published *How to Attract Money Using Mind Power*, a book that set a new standard for concise, no-nonsense, straight-to-the-point self-help books. First published in 2007, that game-changing book continues to transform lives around the world. And though it would be years before James would write new books, and even more years before he would publish new books, that first book set the tone for his writing career. The tagline for James as an author and publisher is Books to Awaken, Uplift, and Empower™. And James takes those words seriously, as is evident in every book he writes. James: is a relative recluse and spends most of his time alone; is an advanced mind-power practitioner, a natural-born astral traveler, and an experienced lucid dreamer; has had life-changing encounters with both angels and demons and even sees some dead people; has been the grateful recipient of an inordinate amount of life-saving divine intervention; is a poet and songwriter; is a genuinely nice guy who cares about people and all forms of life; fasts regularly; is a sincere seeker of higher human health; is an objective observer, a persistent ponderer, and a deliberate deducer; and has a supple sense of heady humor.

STAY IN TOUCH WITH JAMES

If you are a sincere seeker of spiritual truth and/or a determined pursuer of material wealth and success, James could be the lifeline and the go-to resource you have been hoping to find. Step One, subscribe to James's free monthly *Mind Power & Money Ezine* here: jamesgoijr.com/subscriber-page.html. Step Two, connect with James online anywhere and everywhere you can find him. You can start here:

Facebook.com/JamesGoiJr
Facebook.com/JamesGoiJrPublicPage
Facbook.com/HowToAttractMoneyUsingMindPower
Twitter.com/JamesGoiJr
Linkedin.com/in/JamesGoiJr
Pinterest.com/JamesGoiJr
Plus.Google.com/+JamesGoiJr
Youtube.com/JamesGoiJr
Instagram.com/JamesGoiJr
Goodreads.com/JamesGoiJr
jamesgoijr.tumblr.com

James' Amazon Author Page

A great resource to help you keep abreast of James's ever-expanding list of books is his Author Central page at Amazon.com. There you will find all of his published writings and have easy access to them in the various editions in which they will be published. To check out James's page on Amazon, go here:

amazon.com/author/JamesGoiJr

SPECIAL ACKNOWLEDGEMENT

To Kathy Darlene Hunt, who has been my rock, my Light, my safety net, and my buffer since I was in my twenties. She rightfully shares in the credit for every book I've written, for the books I'm working on now, and for every single book I will ever write.

A FREE GIFT FOR YOU!

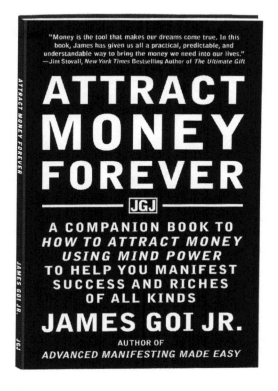

Attract Money Forever will deepen your understanding of metaphysics and mind-power principles as they relate to attracting money, manifesting abundance, and governing material reality. You'll learn how to use time-tested, time-honored, practical, and spiritual techniques to be more prosperous and improve your life in astounding and meaningful ways. Visit jamesgoijr.com/subscriber-page.html for your free download copy of this amazing book and to receive James's free monthly *Mind Power & Money Ezine*.

FURTHER READING

The 80/20 Principle by Richard Koch
The ABCs of Success by Bob Proctor
Abundance Now by Lisa Nichols and Janet Switzer
Act Like a Success, Think Like a Success by Steve Harvey
The Amazing Power of Deliberate Intent by Esther Hicks and Jerry Hicks
As a Man Thinketh by James Allen
The Awakened Millionaire by Joe Vitale
Awaken the Giant Within by Tony Robbins
Being and Vibration by Joseph Rael with Mary Elizabeth Marlow
The Biology of Belief by Bruce H. Lipton, Ph.D.
Breaking the Habit of Being Yourself by Dr. Joe Dispenza
The Charge by Brendon Burchard
Choice Point by Harry Massey and David R. Hamilton, Ph.D.
Clarity by Jamie Smart
The Compound Effect by Darren Hardy
The Cosmic Code by Heinz R. Pagels
The Cosmic Ordering Service by Barbel Mohr
The Council of Light by Danielle Rama Hoffman
Create Your Own Future by Brian Tracy
Creating on Purpose by Anodea Judith and Lion Goodman
Creative Visualization by Shakti Gawain
The Dancing Wu Li Masters by Gary Zukav
The Diamond in Your Pocket by Gangaji
The Dice Game of Shiva by Richard Smoley
Divine Audacity by Linda Martella-Whitsett
The Divine Matrix by Gregg Braden
Dreamed Up Reality by Dr. Bernardo Kastrup

Further Reading

The Dynamic Laws of Prosperity by Catherine Ponder
Emergence by Derek Rydall
Feeling Is the Secret by Neville Goddard
The Field by Lynne McTaggart
Follow Your Passion, Find Your Power by Bob Doyle
The Four Desires by Rod Stryker
Frequency by Penney Peirce
The Game of Life and How to Play It by Florence Scovel Shinn
Having It All by John Assaraf
The Hidden Power by Thomas Troward
How Consciousness Commands Matter by Dr. Larry Farwell
How Successful People Think by John C. Maxwell
I AM by Vivian E. Amis
I Wish I Knew This 20 Years Ago by Justin Perry
Infinite Potential by Lothar Schafer
Instant Motivation by Chantal Burns
It Works by RHJ
Jack Canfield's Key to Living the Law of Attraction by Jack Canfield and D.D. Watkins
Just Ask the Universe by Michael Samuels
Key to Yourself by Venice J. Bloodworth
The Law of Agreement by Tony Burroghs
Lessons in Truth by H. Emilie Cady
Life Power and How to Use It by Elizabeth Towne
Life Visioning by Michael Bernard Beckwith
Live Your Dreams by Les Brown
The Lost Writings of Wu Hsin by Wu Hsin and Roy Melvyn (Translator)
The Magical Approach by Seth, Jane Roberts, and Robert F. Butts
The Magic Lamp by Keith Ellis
The Magic of Believing by Claude M. Bristol
The Magic of Thinking Big by David J. Schwartz

Further Reading

Make Magic of Your Life by T. Thorne Coyle
Manifesting Change by Mike Dooley
The Map by Boni LonnsBurry
The Master Key System by Charles F. Haanel
The Millionaire Mind by Thomas J. Stanley
Mind and Success by W. Ellis Williams
Mind into Matter by Fred Alan Wolf, Ph.D.
Mind Power into the 21st Century by John Kehoe
Miracles by Stuart Wilde
The Miracles in You by Mark Victor Hansen and Ben Carson (Foreword)
Mysticism and the New Physics by Michael Talbot
New Physics and the Mind by Robert Paster
The One Command by Asara Lovejoy
One Mind by Larry Dossey, M.D.
The One Thing by Garry Keller with Jay Papasan
One Simple Idea by Mitch Horowitz
Our Invisible Supply by Frances Larimer Warner
Our Wishes Fulfilled by Dr. Wayne W. Dyer
Physics on the Fringe by Margaret Wertheim
Playing the Quantum Field by Brenda Anderson
The Power of Now by Eckhart Tolle
The Power of Positive Thinking by Dr. Norman Vincent Peale
The Power of Your Subconscious Mind by Joseph Murphy
Power through Constructive Thinking by Emmet Fox
The Power to Get Things Done by Steve Levinson Ph.D. and Chris Cooper
Programming the Universe by Seth Lloyd
Prosperity by Charles Fillmore
Psycho-Cybernetics by Maxwell Maltz
Quantum Creativity by Pamela Meyer
Quantum Reality by Nick Herbert
The Quantum Self by Danah Zohar

Further Reading

Reality Unveiled by Ziad Masri
Reality Creation 101 by Christopher A. Pinckley
The Sacred Six by JB Glossinger
The School of Greatness by Lewis Howes
The Science of Getting Rich by Wallace D. Wattles
The Science of Mind by Ernest Holmes
The Secret by Rhonda Byrne
The Secret of the Ages by Robert Collier
The Self-Aware Universe by Amit Goswami
Shadows of the Mind by Roger Penrose
Shift Your Mind by Steve Chandler
The Slight Edge by Jeff Olson
Soul Purpose by Mark Thurstan, Ph.D.
Spiritual Economics by Eric Butterworth
Supreme Influence by Niurka
There Are No Accidents by Robert E. Hopcke
Think and Grow Rich by Napoleon Hill
Thought Power by Annie Besant
Thoughts Are Things by Prentice Mulford
True Purpose by Tim Kelley
The Universe Is a Dream by Alexander Marchand
Unleash Your Full Potential by James Rick
Warped Passages by Lisa Randall
The Way of Liberation by Adyashanti
What Is Self? by Bernadette Roberts
The Wisdom Within by Dr. Irving Oyle and Susan Jean
Within the Power of Universal Mind by Rochelle Sparrow and Courtney Kane
Working with the Law by Raymond Holliwell
You Are the Universe by Deepak Chopra and Menas C. Kafatos
You Are the World by Jiddu Krishnamurti
Your Invisible Power by Genevieve Behrend
You Unlimited by Norman S. Lunde
The Zigzag Principle by Rich Christiansen